MICHAEL ARTHUR,

UPPERS

Leather and Findings.

A History of Shoemaking

Shoemaking, at its simplest, is the process of making footwear. Whilst the art has now been largely superseded by mass-volume industrial production, for most of history, making shoes was an individual, artisanal affair. 'Shoemakers' or 'cordwainers' (cobblers being those who repair shoes) produce a range of footwear items, including shoes, boots, sandals, clogs and moccasins – from a vast array of materials.

When people started wearing shoes, there were only three main types: open sandals, covered sandals and clog-like footwear. The most basic foot protection, used since ancient times in the Mediterranean area, was the sandal, which consisted of a protective sole, attached to the foot with leather thongs. Similar footwear worn in the Far East was made from plaited grass or palm fronds. In climates that required a full foot covering, a single piece of untanned hide was laced with a thong, providing full protection for the foot, thus forming a complete covering. These were the main two types of footwear, produced all over the globe. The production of wooden shoes was mainly limited to medieval Europe however – made from a single piece of wood, roughly shaped to fit the foot.

A variant of this early European shoe was the clog, which were wooden soles to which a leather upper was attached. The sole and heel were generally made from one piece of maple or ash two inches thick, and a little longer and broader than the desired size of shoe. The outer side of the sole and heel was fashioned with a long chisel-edged implement, called the clogger's knife or stock; while a second implement, called the groover, made a groove around the side of the sole. With the use of a 'hollower', the inner sole's contours were adapted to the shape of the foot. In even colder climates, such designs were adapted with furs wrapped around the feet, and then sandals wrapped over them. The Romans used such footwear to great effect whilst fighting in Northern Europe, and the native Indians developed similar variants with their ubiquitous moccasin.

By the 1600s, leather shoes came in two main types. 'Turn shoes' consisted of one thin flexible sole, which was sewed to the upper while outside in and turned over when completed. This type was used for making slippers and similar shoes. The second type united the upper with an insole, which was subsequently attached to an out-sole with a raised heel. This was the main variety, and was used for most footwear, including standard shoes and riding boots.

Shoemaking became more commercialized in the mid-eighteenth century, as it expanded as a cottage industry. Large warehouses began to stock footwear made by many small manufacturers from the area. Until the nineteenth century, shoemaking was largely a traditional handicraft, but by the century's end, the process had been almost completely mechanized, with production occurring in large factories. Despite the obvious economic gains of mass-production, the factory system produced shoes without the individual differentiation that the traditional shoemaker was able to provide.

The first steps towards mechanisation were taken during the Napoleonic Wars by the English engineer, Marc Brunel. He developed machinery for the mass-production of boots for the soldiers of the British Army. In 1812 he devised a scheme for making nailed-boot-making machinery that automatically fastened soles to uppers by means of metallic pins or nails. With the support of the Duke of York, the shoes were manufactured, and, due to their strength, cheapness, and durability, were introduced for the use of the army. In the same year, the use of screws and staples was patented by Richard Woodman. However, when the war ended in 1815, manual labour became much cheaper again, and the demand for military equipment subsided. As a consequence, Brunel's system was no longer profitable and it soon ceased business.

Similar exigencies at the time of the Crimean War stimulated a renewed interest in methods of mechanization and mass-production, which proved longer lasting. A shoemaker in Leicester, Tomas Crick, patented the design for a riveting machine in 1853. He also introduced the use of steam-powered rolling-machines for hardening leather and cutting-machines, in the mid-1850s. Another important factor in shoemaking's mechanization, was the introduction of the sewing machine in 1846 – a development which revolutionised so many aspects of clothes, footwear and domestic production.

By the late 1850s, the industry was beginning to shift towards the modern factory, mainly in the US and areas of England. A shoe stitching machine was invented by the American Lyman Blake in 1856 and perfected by 1864.

Entering in to partnership with Gordon McKay, his device became known as the McKay stitching machine and was quickly adopted by manufacturers throughout New England. As bottlenecks opened up in the production line due to these innovations, more and more of the manufacturing stages, such as pegging and finishing, became automated. By the 1890s, the process of mechanisation was largely complete.

Traditional shoemakers still exist today, especially in poorer parts of the world, and do continue to create custom shoes. In more economically developed countries however, it is a dying craft. Despite this, the shoemaking profession makes a number of appearances in popular culture, such as in stories about shoemaker's elves (written by the Brothers Grimm in 1806), and the old proverb that 'the shoemaker's children go barefoot.' Chefs and cooks sometimes use the term 'shoemaker' as an insult to others who have prepared sub-standard food, possibly by overcooking, implying that the chef in question has made his or her food as tough as shoe leather or hard leather shoe soles. Similarly, reflecting the trade's humble beginnings, to 'cobble' can mean not only to make or mend shoes, but 'to put together clumsily; or, to bungle.'

As is evident from this short introduction, 'shoemaking' has a long and varied history, starting from a simple means of providing basic respite from the elements, to a fully mechanised and modern, global trade. It is able to provide a fascinating insight not only into fashion, but society, culture and climate more generally. We hope the reader enjoys this book.

THE COLD WATER MAN.

It was an honest fisherman,
    I knew him passing well,
And he lived by a little pond,
    Within a little dell.

To charm the fish he never spoke,
    Although his voice was fine ;
He found the most convenient way
    Was just to drop a line.

And many a gudgeon of the pond,
    If they could speak to-day,
Would own with grief this angler had
    A mighty taking way.

Alas! one day this fisherman
    Had taken too much grog,
And being but a landsman, too,
    He couldn't keep the log.

'Twas all in vain with might and main
    He strove to reach the shore ;
Down, down he went to feed the fish —
    He'd baited oft before.

The jury gave the verdict that
    'Twas nothing else but gin
Had caused the fisherman to be
    So sadly taken in ;

Though one stood out upon a whim,
    And said the angler's slaughter,
To be exac about the fact,
    Was clearly gin and water.

The moral of this mournful tale
    To all is plain and clear—
That drinkers' habits brings a man
    Too often to his bier,

And he who scorns to take the pledge
    And keep the promise fast,
May be in spite of fate a stiff
    Cold water man at last.

TO OUR GERMAN CUSTOMERS.

Write your orders in English if you can, but if you cannot explain yourself in English, then write in German to us.

POSTAL LAW.

MERCHANDISE, consisting of Uppers and small Findings is taken by the Post-Office at one cent per ounce. No package weighing over 4 ℔s. can go by mail, but several packages will be taken to one address.

A Shoemaker taking measures will find this a cheap, prompt and sure method of getting his Uppers and small Findings.

Mention "Ship by Mail" on every order that you wish to go that way, and inclose the price and postage; otherwise, goods not ordered to go by Mail will be sent by Express.

The usual Cost by Mail of Men's Gaiter Uppers is 10c to 15c.
The usual Cost by Mail of Ladies' Gaiters is - - - 7c.
The usual Cost by Mail of Calf or Morocco Legs is - - 25c.

Remember, this is the price to any place in the United States or Territories.

To parties not having a credit with us we send by Express, Cash on Delivery. Small orders should be accompanied by the Money or Post-Office Order, saving return cost of sending money to us. On bills of over $20 we pay return charges.

The first day Artemus Ward entered Toledo, travel-worn and seedy, he said to an elder who was on the street, " Mister, where could I get a square meal for twenty-five cents?" He was told. "I say, mister," said he. "where could I get the twenty-five cents?"

"Patrick, where's the whiskey I gave you to clean the windows with?" "Och, master, I just drank it; and I thought if I breathed on the glass it, would be all the same."

An unsere

Deutschen Kunden.

•———◆◄❮●❯►◆———

Wir erbitten unsere Bestellungen, wenn möglich, in englisch, wer es aber nicht versteht, mag deutsch schreiben. —

Post-Gesetz.

Waaren, welche aus Uppers und kleineren Artikeln bestehen, werden von der Post zu ein Cent per Unze angenommen. Kein Packet darf über vier Pfund wiegen, doch können mehrere Packete an eine Adresse versandt werden. Schuhmacher werden finden, daß dieses ein billiger, prompter und sicherer Weg ist, ihre Uppers zu erhalten. Wenn nicht besonders erwähnt wird, per Post zu senden, so werden alle Packete per Expreß geschickt. Post-Bestellungen müssen übrigens stets mit dem Betrage und Porto begleitet sein.

Das gewöhnliche Porto für Herren Uppers ist.............. 10 Cts.

„   „   „   „ Damen „   „ .... 7 „

„   „   „   „ Kalb oder Morocco Stiefel ist .... 25 „

Dieses gilt für irgend einen Platz in den Vereinigten Staaten.

An Kunden, welche keinen Credit bei uns haben, versenden wir nur per Expreß C. O. D. Für kleinere Bestellungen sollte jedoch stets Geld oder Postanweisung mitgeschickt werden, da auf diese Weise die Retour-Kosten erspart werden.

Für alle Beträge über $20.00 tragen wir Retour-Kosten.

———————

It is better to yield a little than to quarrel a great deal. The habit of standing up, as people call it, for their (little) rights, is one of the most disagreeable and undignified things. Life is too short for the perpetual bickering which attends such a disposition ; and unless a very momentous affair, indeed, where other people's claims and interests are involved, is it not wiser and more dignified to yield somewhat of our precious rights, than squabble to maintain them?

STAND FROM UNDER.

UPPERS REDUCED.

Gent's Button and Congress Gaiter Uppers at 25c. Less.

BOTTOM FIGURES.

We think the custom shoe trade will appreciate the reduction we have just made. Nothing but prompt payment will justify us in selling at such prices. When you get orders for custom work, send to us for the uppers, according to directions in our book. Mention whether to ship by Mail or Express. Buy a Postal Order at your nearest Post-Office. It costs 10 cents for sending $15 or less amount, and 25 cents for a $50 order on New York. Thousands of shoemakers send all their measures to New York and receive them by mail. We send upper and small findings every day to Nevada, Oregon, Florida, Texas, Maine, also to the far-off shores of *New Jersey*. Many of the smaller articles of findings you can receive by mail to advantage if you are in the interior States or Territories. Shoe thread, awls, bristles, silk, machine thread, welts, etc., can go by mail. Send the money or Post-Office orders, including postage, 1 cent per ounce. Please do not send $1 or $2 checks on interior banks. For such small amounts inclose the money or Post-Office order. Banks here charge for collection. Checks for $5 or over are a pleasure to the receiver.

So mote it be!

MICHAEL ARTHUR,

DEALER IN

Leather and Findings, and Upper Manufacturer,

10 SPRUCE STREET,

NEW YORK CITY.

Aufgepaßt!!

Herren

Button und Congress Gaiters,

25 Cents per Paar herabgesetzt!

Die billigsten Preise.

Schuhmacher werden unsere Preisermäßigung anerkennen. Nur prompte Bezahlung setzt uns in den Stand, zu solchen Preisen zu verkaufen.

Für bestellte Arbeit schicke man zu uns für die Uppers nach Angabe in unserer Preisliste. Der sicherste Weg um Geld zu senden ist, eine Post Order zu kaufen. Es kostet 10 Cts. für $15.00 oder darunter, 25 Cts. für $50.00, für Postorders auf New York.

Tausende von Schuhmachern erhalten ihre Uppers von uns per Post. Wir verschicken Uppers und kleinere Artikel täglich nach Nevada, Oregon, Florida, Texas, und allen übrigen Staaten. Manche kleinere Artikel, sowie Zwirn, Ahlen, Borsten, Seide, Maschinenfaden, Rahmen 2c. können vortheilhafter per Post nach inneren Staaten und den Territorien geschickt werden.

Man sende uns entweder das Geld oder eine Post Order, Porto ein Cent per Unze eingeschlossen. Für Beträge von $1.00 oder $2.00 bitten wir unsere Kunden, keine Bankchecks zu schicken, da hiesige Banken für solch kleine Beträge 10 Cts. für collectiren abziehen. Checks für $5.00 und darüber sind uns willkommen.

Michael Arthur,

Leder und Schumacher=Materialien und Verfertiger von Uppers.

10 Spruce St., New-York.

REDUCED 25 CENTS PER PAIR.

GENT'S GAITER UPPERS.
Congress and Button.
ONLY FRENCH CALF USED IN THE VAMPS.

Style No.

No. 26.	Kid top Congress, French Vamps,	- - - -	$2 00
" 29.	Creole Congress, French Vamps,	- - - - -	2 00
" 24..	Broadway Congress, French Vamps,	- - - -	2 00
" 24.	Broadway Congress, Whole Vamps, -	- - - -	2 25
" 14.	Scotch Congress, Whole Vamps, -	- - - -	2 25
" 20.	Seamless top Congress, Whole Vamps,	- - - -	2 25
" 25.	Gent's Button Kid top, -	- - - - -	2 25
" 25.	Gent's Button Kid top, Whole Vamps,	- - - -	2 50

BUCKLE AND LACE SHOES.
ALEXIS REDUCED.

Style No.

No. 15.	Alexis Buckle, Kid tops,	- - - - -	$2 50
" 15.	Alexis Buckle, Calf or Morocco tops,	- - - -	2 40
" 16.	Alexis Lace, 3 eyelets, Calf tops, -	- - - -	2 15
" 3.	English Waukenfasts, with hooks,	- - - -	2 25
" 28.	Foxed Balmorals, Calf top	- - - - -	2 00
" 12.	Webster Ties -	- - - - -	1 75
" 4.	Instep Buckle Shoe, -	- - - - -	2 00
" 21.	Congress Shoe, fancy instep,	- - - -	2 00
" 31.	Congress Shoe, all one piece,	- - - -	2 15
" 23.	Oxford Tie, French, ($1 35) Whole Vamps,	- - -	1 60
	Cloth Overgaiters, no seam,	- - - -	1 00

The above Gaiters are on hand ready fitted; if our customers want a pair made special for any reason, whether to measure or not, it will cost 25 cents extra for each pair so made.

Cutting to Measure, - 25 cents.	Imitation Tips, - - 15 cents.
Corded Wrinkles, - 15 cents.	Patent Lea. Gaiters, - 50 cents extra.
Calf Linings in Custom	Alligator Foxings, $1 50 extra.
Gaiters, - - - 15 cents.	

Send a scrap of Calf to show substance wanted.

SAMSON AND THE JAW-BONE.—To illustrate how curiously persons sometimes try to explain matters that are a hard task for our credulity, I mention a little incident experienced by the writer of these lines. When I traveled, in 1871, in Palestine, an old gray friar from the monastery of Ramieh, about fifty miles west from Jerusalem, showed me the supposed place where Samson killed 1,000 Philistines with the jawbone of an ass. When I expressed my doubts as to the length and strength of a jawbone, considering the great number of surrounding enemies, the good monk explained the case in the following manner: "Well, he took hold of the ass by the tail and swung the animal against the Philistines in such a manner that only his head, and of this especially the jaw-bone, struck the Philistines, keeping off in this way the surrounding warriors, and giving the blow the necessary force to kill." I affirm that in this manner Samson could have slain a million Philistines, provided the tail of the ass did not break.—*Sacramento Journal.*

EXTRA QUALITY GAITERS.

We find that many of our customers are calling for extra good gaiters with vamps selected from Mercier and Ulmo French calf. We now keep the following kinds of uppers with the vamps stamped " extra " gaiters, and at the following prices:

Style No.

No. 26.   Kid top Congress,   -   -   -   -   -   -   - $2 25
 " 29.   Creole Congress,   -   -   -   -   -   -   -   -   2 25
 " 20.   Seamless top Whole Vamps Congress,   -   -   -   2 50
 " 25.   Gent's Button Whole Vamps,   -   -   -   -   -   2 75
 " 25.   Gent's Button Kid top, -   -   -   -   -   -   -   2 50

Cutting to measure 25 cents extra.

Unless the order calls for "EXTRA " gaiters we shall not send these.

Send a scrap of calf to show the substance wanted.

***

TIGHT BOOTS.

I would like to kno who the man waz who fust invented tight boots.

He must hav bin a narrow and kontracted kuss. If he still lives I hope he haz repented ov his sin, or iz enjoying grate agony ov sum kind.

Enny man who kan wear a pair of tight boots and be humble and penitent and not swear will mak a good husband.

Mi feet are az uneasy az a dog's nose the fust time he wears a new muzzle.

I think mi feet will eventually choke the boots to deth. I liv in hopes they will.

Avoid tight boots my friend az you would the grip of the devil, as many a man learns to swear by encouraging hiz feet to hurt hiz boots.

I am too old and too respectable to be a phool enny more. Tite boots are an insult to enny man's understanding. He who wears tite boots will hav to acknowledge the corn. Tite boots hav no bowels ov mersy, their insides are wrath and promiskious cussin.

Beware ov tite boots.

LOOK HERE!
CHEAPER UPPERS.

We are selling a well cut and fitted Gaiter Upper, with American Calf tops and vamps, at these prices, which will enable any Shoemaker to make up Gaiters ahead, to fill in any idle time of his hands or himself.

These are not Rubbish. They will look well and wear well.

Style 26.	Calf Congress Men's Uppers - - - - -	$1 00
" 26.	Calf Congress " " better article - - -	1 25
" 14.	Scotch Congress Whole Vamps - - - -	1 50
" 25.	Gent's Button Calf Top - - - - - -	1 50
" 25.	Gent's Button Whole Vamp - - - - -	1 75
" 15.	Alexis Buckle - - - - - - - -	1 25
" 29.	Creole Congress - - - - - - -	1 75
" 3.	English Waukenfast Hooks - - - - -	1 75
" 12.	Webster Tie - - - - - - - -	1 00
" 23.	Oxford Tie - $1 00 Instep Buckle, - - -	1 25

THESE GOODS WE DON'T CUT TO CUSTOM MEASURE.

Try a Pair as Sample, then You will Buy More.
BOYS' WORK.

Boys' Congress Gaiters, 1 to 5, American Calf - - - -	$1 00
Boys' Calf Top Button, 1 to 5, " " - - - -	1 35
Boys' Balmorals, 1 to 5, - - $1 00 French Kip, best - -	1 30
Youths' Balmorals, 8 to 13 - 90 " " best - -	1 15
Boys' Kip and Calf Boot Legs, 1 to 5 - - - - - -	2 90

WIT.—"There is no more interesting spectacle than to see the effect of wit upon the different characters of men; than to observe it expanding caution, relaxing dignity, unfreezing coldness, teaching age, and care, and pain to smile, extorting reluctant gleams of pleasure from melancholy, and charming even the pangs of grief. It is pleasant to observe how it penetrates through the coldness and awkwardness of society, gradually bringing men nearer together, and like the combined force of wine and oil, giving every man a glad heart and a shining countenance. Genuine and innocent wit like this is surely the flavor of the mind! Man could direct his ways by plain reason, and support his life by tasteless food; but God has given us wit and flavor, and brightness, and laughter, and perfumes, to enliven the days of man's pilgrimage, and to ' charm' his pained steps over the burning marl.'"—*Sydney Smith.*

BOOT LEGS.

No. 6.	French Calf Boot Legs, - - - - - - -	$3 75
" 6.	French Calf Legs, short fronts, 2d quality, - - -	3 00
" 6.	French Kip, or very stout Calf Legs, - - - -	4 00
" 6.	French Calf Legs, from extra Brands of Stock, - -	4 25
" 11.	Light Grain, side seam Leg, French Tongue feet, - -	3 50
" 11.	Mor. Leg, French Calf feet, - - - - - -	4 25
" 11.	Mor. Leg, French Patent Leather feet, - - - -	5 00
" 11.	Mor. Leg, Alligator feet, - - - - - -	7 00
" 1.	Napoleon Legs, all Grain, Tongue feet, - - - -	7 00
" 1.	Napoleon Legs, Grain, French Kip feet, - - -	7 50
.	Grain Hunting Legs, Kip or Grain feet, - - - -	9 50
	Grain Riding Legs, - - - - - - -	8 50
" 1.	Napoleon Enamel Leather and French Kip feet - -	8 50
" 2.	Napoleon Leggings, Enamel Leather or Grain - - .	4 50

Boots cut to measure, 25 cents extra.

Footing legs, - - - per dozen, 10 00

OUR UPPERS MEASURE IN THE HEEL:

5	6	7	8	9	10	11
12	12½	12¾	13	13½	14	14¼ heel.

☞ *Send scrap to show substance wanted.*

A paper encourages the young by the example of a "youth who formerly lived in a hovel; yet with only his two hands and a crow-bar opened a jewelry store; and now he lives in a large stone residence in Sing Sing."

A smart young man, so fresh that his parents think of having him rubbed down with rock salt, attempted to run the elevator at an uptown hotel the other day while the operator was flirting around the corridor with a red-headed house maid. He pulled the rope and up started the car, and the smart young man looked out and shouted to the bell boys: "How's this?" Just as he shouted the back of his head came in violent contact with the woodwork of the second-floor landing, and he bit his tongue and saw fireworks simultaneously. The elevator was stopped, and they took him out. He had claret on his lips, and a patented carbuncle on the back of his head. "Fellows," said he, making a violent effort to smile, "fellows, do any of you know anything new in profanity?"

A man stepped on a piece of banana peel on Main street this morning, and promptly sat down on a pint bottle of yeast in his coat pocket. He rose immediately.—*Bridgeport Standard.*

LADIES' FRENCH GLAZED KID.

Cut from Bassett—Extra Choice.

Style No.
No. 5. French Glazed kid, Button, with Vamp, - - - $3 00
" 9. French Glazed Kid, Button, seam from top to toe, - - 3 00

LADIES' FRENCH CALF KID.

No. 5. Calf Kid, Button, genuine French stock, - - - - 2 50
" 7. Calf Kid, Po. Lace, - - - - - - - - 2 00
" 8. Calf Kid, Congress, new pattern, - - - - - - 2 25

Cutting to measure, 25 cents extra.

LADIES' FRENCH OIL MOROCCO.

No. 5. Oil Morocco, Button, imported Stock, - - - - 2 15
" 7. Oil Morocco, Lace, - - - - - - - 1 65

LADIES' LASTING OR SERGE.

No. 2. Lasting, Congress, Serge, - - - - - - - 1 25
" 5. Lasting, Button, - - - - - - - - 1 65
" 7. Lasting, Laced, - - - - - - - - 1 15

LADIES' CALF.

No. 7. Ladies' Calf, Po. Lace, - - - - - - - 1 90

MISSES' BUTTON AND LACE.

No. 33. Misses' French Glazed Kid, Button, 8 to 1, - - - 2 00
" 33. French Oil Morocco, Button, 8 to 1, - - - - 1 60

Morocco Lace, 25 cents less.

ON HORNS.

Horns varys in length, but from three tu six inches iz the favorite size. It is different from other horns being ov a fluid natur. It iz really more pugnashus than the ram's horn; six inches ov it will knok a man perfekly calm. When it knocks a man down it holds him there. It has drawn more tears, broken more hearts and blited more hopes than all the other agencys ov the devil; what kind ov pizen will you take stranger?

The man who is only honest because he thinks honesty is the best policy, is not really an honest man. Honesty is not swerving policy, but stable principle. An honest man is honest from his inmost soul, nor deigns to stoop to aught that is mean, though great results hang on the petty fraud.

HER SANGUINE TEMPERAMENT.

The other evening the police were informed that a resident of Lafayette street East was killing his wife. This is not an unusual thing for a husband to do, during these days of pull-backs, tie-backs, get-backs, back-ups, and long trails, but still the policeman made a rush for the house. As he reached the steps the wife had just finished washing her bloody nose, and she greeted him with the cheerful query:

"Hello! Did you hear of the racket?"

"I heard that you were being murdered," he replied.

"Oh! pshaw! It was merely a lively little set-to between the old man and myself. We have lots of 'em. I don't always come out second best, as I did this time, but it's all right."

"If a body meet a body
Comin' through the rye!"

"I should think it would be awful to live this way," remarked the officer, as he glanced at the many proofs of poverty.

"Oh, go 'long," she smiled; "we can't all be dukes and dukesses, and there's no use trying. I've got six children around the house, and it's my duty to carry a lively heart. Fact is, I'm of a sanguine temperament, and I always look on the bright side anyhow."

"Weren't you set out of a house on Croghan street for non-payment of rent?" the officer asked, looking at her more closely.

"Same woman—same family—same family," she laughed; "I had more fun over that than you could carry on a freight train. Three of the children were sick, the old man out of work, the dog lost, the cat under the weather with the cramps, and none of us knew what to do. However,

'The sun may be shining to-morrow,
Although it is cloudy to-day;'

and I sat down on the old cook stove and laughed till I cried."

"I think I saw you at the poormaster's office," he observed.

"And that was another good joke on Snyder," she grinned. "Yes, I went around there and asked for Mocha coffee, granulated sugar, seedless raisins, Worcestershire sauce, pastry flour, and A 1 coal, and you ought to see the old man go down in his boots. I got some taters, and meat and wood, and some of the old folks were put out to hear me singing—

'The wolf of starvation, she winked at me—by-by—tra-la!
But I married a duke with a fortune three—fe—fo—fum!'"

"Well, I guess you'll get along," said the officer, as he went down the steps.

"Don't you bet I won't!" she replied, standing in the door. "We haven't a stick of wood, and nothing to eat but a loaf of bread, while the rent is two months overdue, but I am of a sanguine temperament, you know. If we don't strike a streak of luck to-night we'll have a dry old meal and another fight in the morning, but luck has got to come some day. Destiny is destiny, and this old calico dress has got to do me till snow flies, but—

'There's many a hard up fam-i-lee—there's many who want for bread;
But I'm a sandy, sanguine, cheerful wife; who'll never give up till dead.'"

"If you hear a tussel in here this evening don't interfere. I've got a handful of snuff all ready for the old man's eyes, and it'll nearly kill me to see him fooling around for a club with one hand and digging his eyes with the other. Well, tra-la."—*Detroit Free Press.*

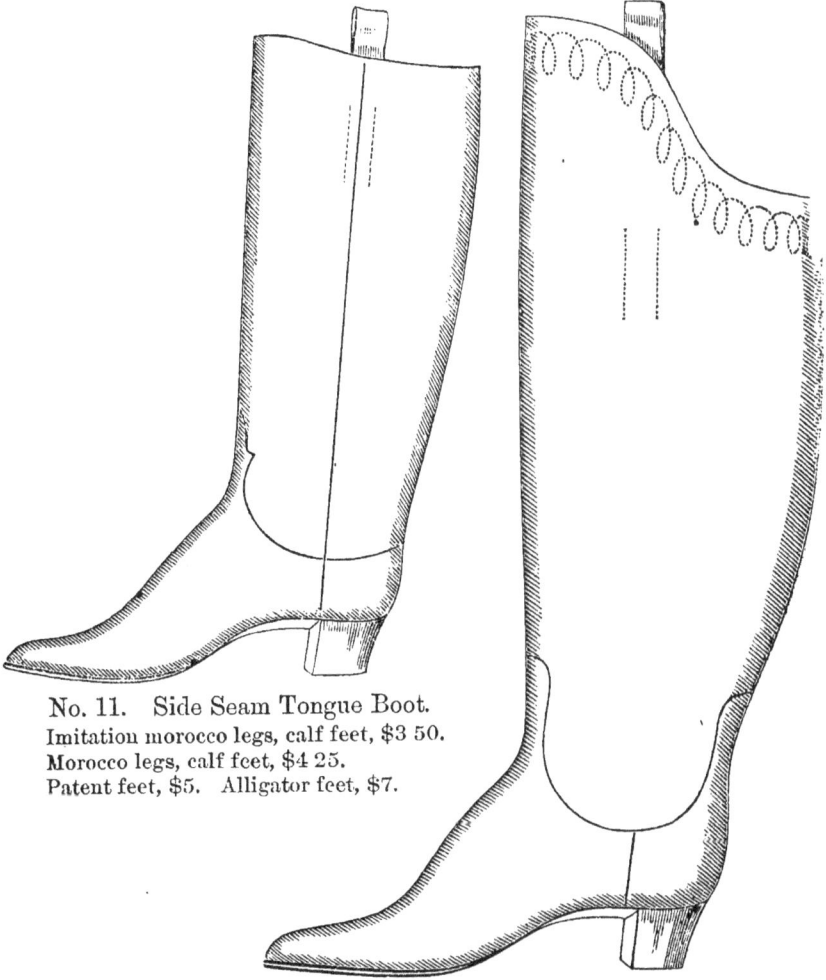

No. 11. Side Seam Tongue Boot.
Imitation morocco legs, calf feet, $3 50.
Morocco legs, calf feet, $4 25.
Patent feet, $5. Alligator feet, $7.

No. 1. Napoleon.
All grain, $7. French kip feet, $7 50.

"Pat, buy a trunk to put your clothes in," said his Yankee companion. "What! and go naked this cold weather?" asked the honest spalpeen of Killarney.

Jones told Brown that he could always get away from home of an evening by telling his wife that the noise of her sewing machine made him nervous, and that nothing composed him so well as a long rattling walk; but when Brown tried this dodge on his wife she kicked his hat under the sofa and said, "You can't play any of Bill Jones' tricks on me, and this Light Running 'Domestic' doesn't make any noise, either. I ain't permanently insane if I did marry you." Brown didn't go out that night, but he would have given a week's salary to have been somewhere else when she found out that he had lighted his cigar with her 'Domestic' paper fashion for a winter dolman.

No. 25.   Gent's Button.
$1 50. $2 25. $2 50.

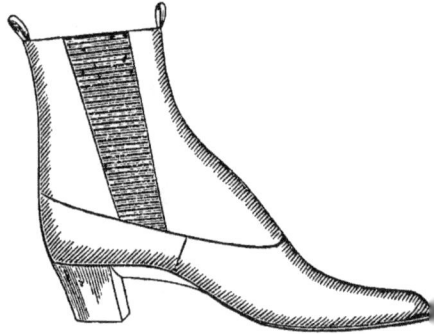

No. 26.   Kid Top Congress.
$1. $1 25. $2. $2 25.

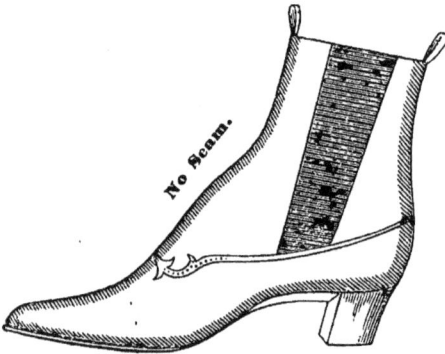

No. 14.   Scotch Congress.
$2 25. $2 50.
Whole Vamp.

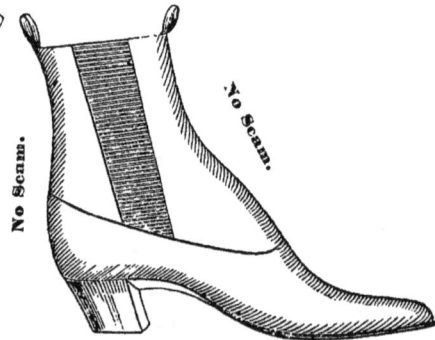

No. 20.   Calf Top Congress.
$2 25. $2 50.
Seamless Top, Whole Vamp.

During the strike in Albany, while Coroner Fitzhenry, of that city, who is a member of the Burgesses Corps, was guarding the western end of the upper railroad bridge, a man attempted to pass the guard. The coroner commanded the intruder to halt. "Who will stop me from going over this bridge?" asked the man. "I will," said the coroner. "Would you stop the likes of me, who voted for you for coroner?" The coroner replied, "I am put here to shoot, and I get thirty dollars for a corpse. If you don't leave I'll put a bullet through you."

An angry letter never accomplishes the desired end, and an insolent one harms none but the writer. This is true of all correspondence, but more especially when applied to communications of a business nature. In this department the true gentleman is easily recognized, and with him, above all others, it is gratifying to deal. His demands, which if couched in other language would be rejected, are often complied with, and, whatever the business, there is satisfaction in performing it.

No. 31. Congress Shoe. $2 15.
All one piece.   A Handsome Shoe.

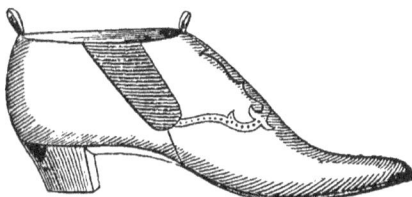

No. 24. Broadway.
$2 25, $2 50; Pieced, $2.
Whole Vamp. New Pattern.

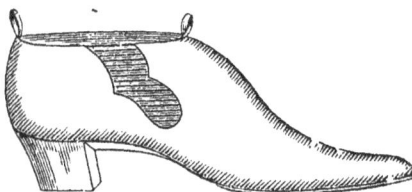

No. 21. Congress Shoe. $2.
Fancy Vamp.

No. 16. Alexis Lace.
$1 25, $2 15.

No. 15. Alexis Buckle.
$1 25, $2 40, $2 50.

HOW THEY WALTZ IN PUT-IN-BAY.

People may say that a waltz is a waltz, but it is a mistake; as much as to say that a dog is a dog; for there are dogs and dogs, and there are waltzes and waltzes. With one person it is the poetry of motion; with another it is about as awkward a performance as putting yourself upon a level and going through the motion of running up stairs would be. A Kentucky girl is a natural waltzer, and she does it with a charming *chic* and *abandon*. An Ohio girl's waltzing is easy, graceful and "melodious." If she happens to come from Cincinnati and across the Rhine, she swings dreamily round and round in the endless "Dutch waltz." If she comes from Chicago, she throws her hair back, jumps up and cracks her heels together, and carries off her astonished partner as though a simoon had struck him, and knocks over all intervening obstacles in her mad career around the room. If she is from Indiana, she creeps closely and timidly up to her partner as though she would like to get into his vest pocket, and melts away with ecstacy as the witching strains of the "Blue Danube" sweep through the hall. If she is from Missouri, she crooks her body in the middle like a door hinge, takes her partner by the shoulders and makes him miserable in trying to hop around her without treading on her No. 9 shoes. If she comes from Michigan, she astonishes her partner by now and then working in a touch of the double-shuffle, or a bit of pigeon-wing, with the waltz step; and if she comes from Arkansas, she throws both arms around his neck, rolls up her eyes as she floats away, and is heard to murmur, "Oh, hug me, John!"

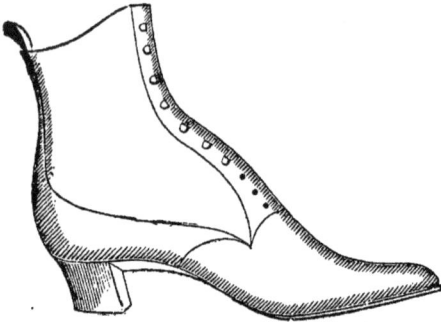

No. 3. Waukenfast, (Hooks).
$1 75, $2 25.

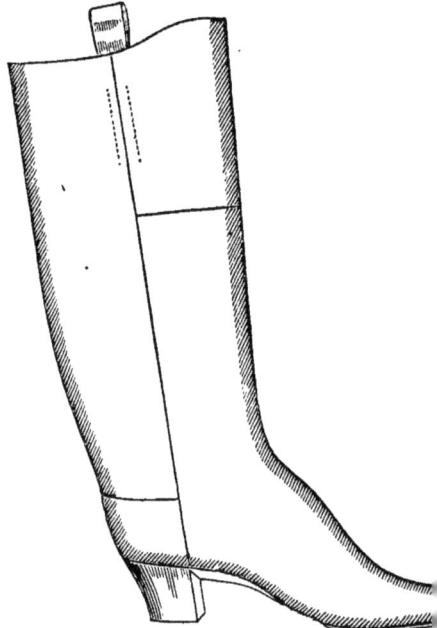

No. 29. Calf Creole.
$1 75, $2, $2 25.

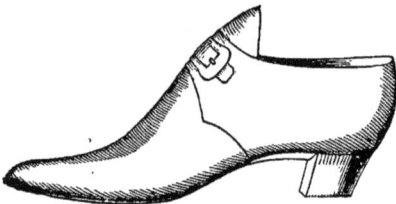

No. 6. Calf Boot Leg.
$3 75, $4 25.

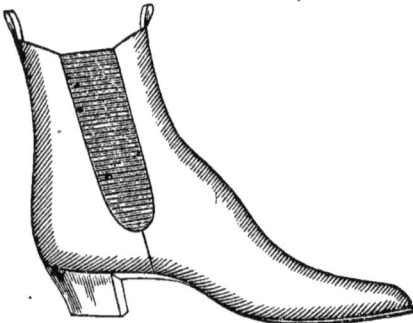

No. 4. Instep Buckle. $2

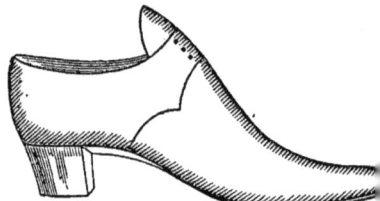

No. 12. Webster Tie. $1 75.

A Providence deacon, a few evenings since, at a prayer meeting in that city, arose and e
pressed himself as follows: "My friends, with great sorrow and regret I have just learn
of the decease of our beloved Brother ———. Let us now sing ' Praise God from whom a
blessings flow.'"

An Irishman's friend having fallen into a slough, the Irishman called londly to another f
assistance. The latter, who was busily engaged in cutting a log, and wished to procrastinat
inquired, "How deep is the gentleman in?" "Up to his ankles." "Then there is plenty
time," said the other. "No, there is not," rejoined the first; "I forgot to tell you he's
head first."

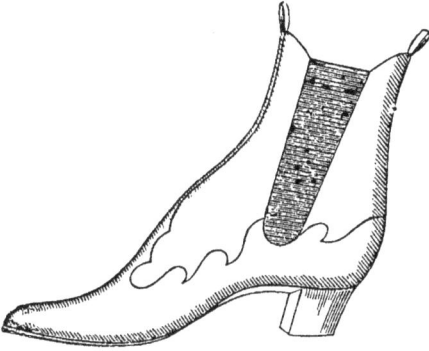

No. 8.   Ladies' Congress.   Calf Kid. $2 25.

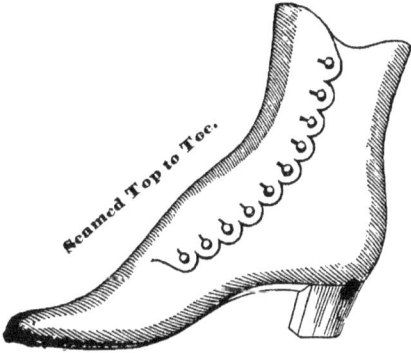

Seamed Top to Toe.

No. 9.   Ladies' Button.
Kenny & McPartland's Patent, New York.

No. 5.   Ladies' Button.

A little boy who was nearly starved by a stingy uncle, with whom he lived, meeting a lank greyhound one day in the street, was asked by his guardian what made the dog so thin. After reflecting, the little boy replied, "I suppose he lives with his uncle."

A merchant, advertising for a clerk who could bear confinement, received an answer from one·who had been seven years in jail.

PADDY SHANNON.

Paddy Shannon was a bugler in the Eighty-seventh Regiment—the Faugha-Ballaghs—and, with that regiment under the command of Sir Hugh Gough, served all through the Peninsular campaign. When the campaign was over, Paddy had nothing left him but the recollections of it. His only solace was the notice taken of him in the canteen. It is no wonder, then, he became a convivial soul. From the bottle he soon found his way to the halberts.

The regiment was paraded, and the proceedings read, and Paddy tied up. The signal was given for the drummers to begin, when Paddy Shannon exclaimed : " Listen, now, Sir Hugh. Do you mean to say you are going to flog me ? Just recollect who it was sounded the charge at Borossa, when you took the only French eagle ever taken. Wasn't it Paddy Shannon ? Little I thought that day it would come to this ; and the regiment so proud of that same eagle on the colors." "Take him down," said Sir Hugh, and Paddy escaped unpunished. A very short time, however, elapsed before Paddy again found himself placed in similar circumstances. "Go on," said the Colonel. "Don't be in a hurry," ejaculated Paddy. "I've a few words to say, Sir Hugh." "The eagle won't save you this time, Sir." "Is it the eagle, indeed ! then I wasn't going to say any thing about that same, though you are, and ought to be proud of it. But I was just going to ask if it wasn't Paddy Shannon who, when the breach of Tarifa was stormed by 22,000 French, and only the Eighty-seventh to defend it, if it wasn't Paddy Shannon who struck up, ' Garryowen, to glory, boys,' and you, Sir Hugh, have got the same two towers, and the breach between them, upon your coat of arms in testimony thereof." "Take him down," said the Colonel, and Paddy was again unscathed.

Paddy, however, had a long list of services to get through, and a good deal of whiskey, and ere another two months he was again tied up, the sentence read, and an assurance from Sir Hugh Gough that nothing would make him relent. Paddy tried the eagle ; it was of no use. He appealed to Sir Hugh's pride and the breach of Tarifa without any avail. "And is it me," at last he broke out, "that you are going to flog? I ask you, Sir Hugh Gough, before the whole regiment, who kn w it well, if it wasn't Paddy Shannon who picked up the French Field Marshal's staff at the battle of Vittoria, that the Duke of Wellington sent to the Prince Regent, and for which he got that letter that will be long remembered, and that made him a Field-Marshal into the bargain? The Prince Regent said : ' You've sent me the staff of a Field-Marshal of France; I return you that of a Field-Marshal of England.' Wasn't it Paddy Shannon that took it? Paddy Shannon, who never got rap, or recompense, or ribbon, or star, or coat-of-arms, or mark of distinction, except the flogging you are going to give him ?" "Take him down," cried Sir Hugh, and again Paddy was forgiven.

PROPER WAY OF ORDERING CUSTOM UPPERS.

Send the Money for Uppers and Postage.

We have an eminent respect for Cash transactions.

We could do such a thing as charge an account to very good parties, but recognize that the world is divided very much as to who such parties are.

" Short accounts make long friends."

EXTRAS.

Cutting to measure - - - 25 cents.	Calf Linings - - - - 15 cents.
Corded Wrinkles - - - 15 "	Patent Leather Gaiters - 50 extra.
Imitation Tips - - - - 15 "	Alligator Foxings, - - $1 50 "

Men's Extra Gaiters, when made special, 25c. more.

Send one pair Kid Top Congress Extra Gaiters, Fig. 26, Calf Lined ; Sample inclosed. Send by Mail.

Last	Heel	Instep	Ball	Small Ankle
9	14	$9\frac{1}{2}$	$9\frac{1}{2}$	10

Uppers, $2 25. Cutting, 25c. Calf Lining, 15c. Postage, 10c. Inclosed, $2 75.

Send one pair of Gent's Kid Top Button Gaiters. Fig. 25, medium substance like sample, wrinkled.

Last	Heel	Instep	Ball	Small Ankle
9	$13\frac{1}{2}$	10	9	$9\frac{1}{2}$

Uppers, $2 25. Cutting, 25c. Wrinkles, 15c. Postage, 10c. Inclosed, $2 75.

Send one pair Calf Legs (No. on Card 6), like substance inclosed.

Last	Heel	Instep	Ball	Calf of Boot
10	14	$9\frac{1}{2}$	$9\frac{1}{2}$	$15\frac{1}{4}$

Uppers, $3 75. Cutting, 25c. Postage, 25c. Inclosed, $4 25.

Send one pair Ladies' French Glazed Kid Button, Black Stitch, (No. on Card 5).

Last	Heel	Instep	Ball	Small Ankle	Top	High
3	$11\frac{1}{2}$	$8\frac{1}{4}$	$8\frac{1}{2}$	8	10	$7\frac{1}{2}$

Uppers, $3. Cutting, 25c. Postage, 7c. Inclosed, $3 42.

Please be particular in describing the kind of stock wanted.

The cheapest way to get a pair of Uppers is to have them sent by Mail. Prepay and order that way.

WHERE TO MEASURE WHEN YOU ORDER UPPERS FROM US

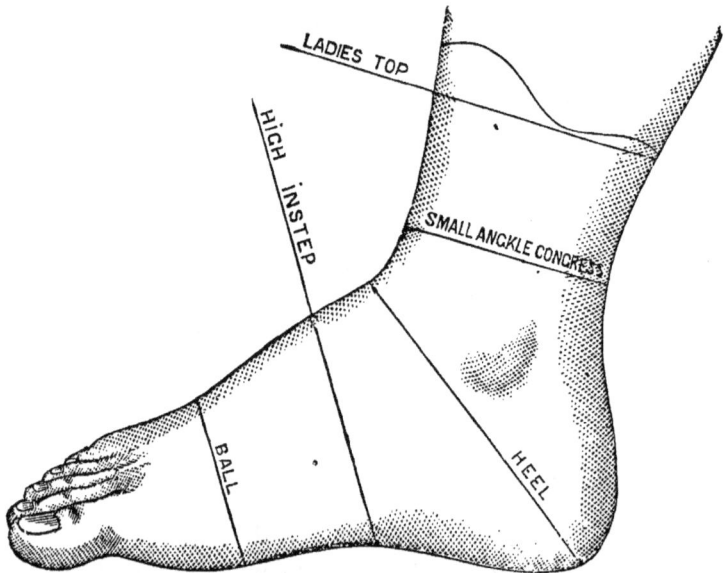

PATTERNS.

We sell paper copies of our patterns, all of which we have lately improved.

Any 1 set, all sizes - - - - - - - - - $3 00

" 5 " " " - - - - - - - - - 10 00

For instance, Boot style No. 6 has 44 pieces ; Gent's Gaiters style No. 26 has 40 pieces, etc.

No one can cut Custom patterns accurately and quickly unless they have regular sets to go by.

Money laid out in good patterns is well spent.

NEW ELASTICS.

New Elastics inserted in ready-made Gaiters 65c. for Common and **75c.** for silk Terry.

A Discount made for large quantities.

———

He was from Nebraska. He was passing by a White street fish market when he saw a lobster on a bench. The sight was so unexpected that he lost his presence of mind, and before he could recover himself had openly confessed. " By jinks, that's the biggest grasshopper I have ever seen."—*Danbury News.*

An ingenious Frenchman on Long Island claims to have discovered a sure means of destroying the potato bugs. Mix one gallon of prussic acid with three ounces of rend-rock, stir well, and administer a tablespoonful every hour and a half till the bug shows signs of weakening. Then stamp on him.

FRENCH CALF AND FRENCH KIP BOOT FRONTS AND FOOTINGS, CRIMPED.

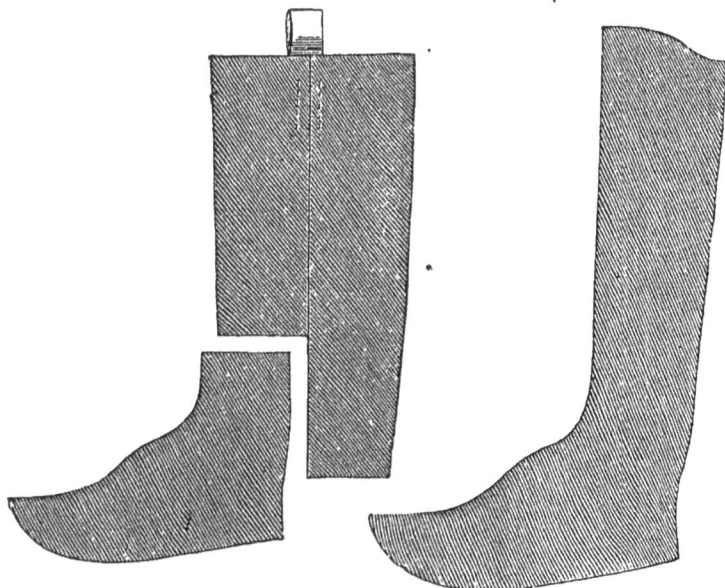

French Calf Skin Boot Fronts, - - - - - - - $1 75
French Calf Skin Footings, - - - - - - - - 1 25.
French Fronts, from Extra Brands, Ulmo and Mercier, - - 2 00
French Kip Fronts, - - - - - - - - - 2 00
French Kip Footings, - - - - - - - - 1 50
American Oak Calf Fronts, - - - - - - - - 1 50
"   "   "   " Footings, - - - - - - 1 00
American Oak Kip Fronts, - - - - - - - - 1 50
Boot Backs, Calf Skin, - - - - - - 50 cents to 90
Wax Oak Kip Backs, high cut, - - - - - 75 " " 1 00

Send a small scrap for substance.

HOW TO PICK A WATERMELON.—Some time late in August when the moon is up and the clock has struck twelve, midnight, get up and dress, without any noise, go out the back way and turn to the right, after going a half mile go to the left, over a fence, and you will see the melon patch. Pick out a dark colored one, with the skin a little rough, shoulder the melon and step out briskly ; once and a while look over your shoulder to see if the moon is all right. When you are home bury the melon in the hay mow, and slip into bed. This is an old fashioned way to pick melons, and the way your father and grandfather picked them. After you get the melon tear up the receipt. One melon during your life is enough to pick in this way.

BOTTOM STOCK.

Hemlock Sole Leather, best, 20 to 29 ℔s, - - - - - 26c.

Hemlock Sole Leather, best, 14 to 19¾ ℔s, - - - - 25c.

Hemlock Sole, medium quality, - - - - - - - 24c.

Hemlock Good Damaged, - - - - - - - 22c.

OAK LEATHER.

Baltimore Oak Sole, - - - - - - - - - 36c.

Philadelphia Oak Sole, - - - - - - - - 35c.

Hoyt's Flintstone Oak Sole, no bellies or heads, - - - 45c.

Hoyt's Flintstone Cuts, all soles cheap and prime.

Broad Oak Bellies, for insoles, - - - - - - - 22c.

Union Crop Leather, extra choice, no heads, - - - 37c.

CUT BOTTOM STOCK.

Hoyt's Flintstone Oak Soles, per dozen pair, - - $4 00 to $7 50

Hoyt's Flintstone Half Soles, - - - - 3 00 to 4 50

Hemlock Half Soles, - - - - - - - 2 50 to 3 50

Inner Soles, - - - - Per dozen pairs, 2 25

All the above stock is square cut.

"Sire, one word," said a soldier one day to Frederick the Great, when presenting to him a request for the brevet of Lieutenant. "If you say two," answered the King, "I will have you hanged." "Sign," replied the soldier. The King stared, whistled, and signed.

His reading the following advertisement in a New Haven paper was the origin of the rumor that the Postmaster was going to resign : "Any person having five to fifty loads of manure to dispose of will please send word or drop it through the Post-Office."

FRENCH CALF SKINS.

As a very desirable French Calf Skin, with good flanks and shoulders, and a reputation for wear equal to anything in the market, we can recommend the Ulmo Brand for very fine custom trade.

	LBS. PER DOZ.	PER LB.
French Calf, Simon Ulmo, - - - - -	22 to 33	$1 80
French Calf, Simon Ulmo, - - - - -	36 to 38	1 70
French Calf, Simon Ulmo, - - - - -	40 to 44	1 52
French Calf, Ulmo, Females, - - - -	22 to 33	2 10
French Calf, Ulmo, Mixed, Choice, nice goods, - -	22 to 33	1 56
French Calf, Ulmo, Mixed, Choice, - - -	38	1 52
French Calf, Leven, - - - - - -	22 to 32	1 85
French Calf, Leven, - - - - - -	35 to 38	1 75
French Calf, Corneillan, - - - - .- -	30 to 33	1 85
French Calf, Corneillan, - - - - -	35 to 37	1 75

	10 KILO.	11 KILO.	12 KILO.
French Calf, Jodot Firsts, - - -	$58 00	$64 00	$73 00
Swiss Calf, Mercier, - - - - -		23 to 29	2 15
Swiss Calf, Mercier, - - - - -		31 to 34	2 10
Swiss Calf, Mercier, - - - - -		36 to 38	1 90
French Calf, Edward Dietz, - - - -		26 to 32	1 50
French Calf, Edward Dietz, - - - -		33 to 38	1 40
French Calf, Edward Dietz, - - - -		40 to 43	1 35
French Calf, Edward Dietz, - - - -		44 to 48	1 25
French Calf, Simon, - - - - -		26 to 32	1 50
French Calf, Simon, - - - - -		35 to 40	1 45

☞ On any less quantity than half dozen we charge 10 cents per ℔ extra.

FRENCH KIP SKINS.

Reichlen French Kips, - - - 9 lbs. per skin, 85 cents per ℔.
Koch French Kips, - - - - 7 lbs. per skin, $1 05 " "

AMERICAN STOCK.

American Calf, Oak Tan, ordinary, - - - - $ 90 to 1 00
American Calf, Oak Tan, best, - - - - 1 15
American Calf, Hemlock Tan, - - - - · - 80 to 90
American Kip, - - - - - 50 to 70 ℔s., 85
American Kip, - - - - - 73 to 80 ℔s., 75
Wax Upper, - - - - - Per foot, 21
Oak Splits, - - - - - - Per ℔., 45
Welt Leather, prime selection, - - Per side, 4 00 to 6 00
Welt Leather, cut in dozen pairs, - - Per doz., 40 to 1 00
Smooth Grain and Buff Leather, - Per foot, 20
Grain Leather, for waterproof boots, - Per side, 5 00 to 6 00
Alligator Skins, - - - - Each, 4 00 to 6 00
Buckskins, - - - - - - Each, 3 00 to 5 00
Best French Patent Leather, silver medal, Per skin, 3 50

SHEEP SKINS.

	No. X.	XX.	XXX.
Cream Sheep, New York, - - -	$6 75	$8 00	$9 00
Common Creams, - - - -	4 50		
Colored, for topping, extra goods, -	8 00	9 00	
Pink Linings, - - - -	5 50	6 50	7 50
Bark Linings, - - - - -	4 50	6 00	7 00
Alum, Soft White, for ladies' uppers, imported,			8 50
Apron Skins, - - - - - - -		Each, 75c. to 1 00	

FRENCH GLAZED KID.

	XXX.	
Grison, extra choice, - - - - -	$26	Per skin, $2 25
Bassett, extra choice, - - - -	26	Per skin, 2 25

CALF KID SKINS.

Selected Calf Kid, prime goods, - - - - - 25c. per foot.

Try these goods.

MOROCCO AND GOAT SKINS.

Tampico Boot Morocco, black, - - - -	$2 25 to $3 00 per skin.
Tampico Boot Morocco, maroon - - -	3 00 to 3 75 "
Smooth Glazed Morocco, - - - - -	1 50 to 2 25 "
American Oil Morocco, Pebble or Long Grain,	2 25 to 3 00 "
French Long Grain Morocco (genuine), per doz., 32 00	Per skin, 3 00

If possible send sample in ordering.

TERRY ELASTICS.

Wool, Terry, new article, 5 and 5½ inches,	Per yd., 60 cts.	65
Silk, Terry, Goring, 5 inches, - - -	" 80 "	$1 10
Silk, Terry, Goring, 5½ " - - - -	" 85 "	1 15
Cotton, Terry, 5 " - - -	40 cts. " 45 "	
Ladies' Silk, Terry, 5½ " - - - - -	Per yd. 1 10	

LASTS.

Men's and Women's Fine Maple Lasts, - - - - -	40 cts.
Men's and Women's Straight Block, - - - - -	21 "
Youths' and Misses' Straight Block, - - - - -	18 "
Youths' and Misses', Right and Left, - - - -	36 "
Boys' Lasts, - - - - - - - - -	40 "

BOOT TREES.

	Plain.	With Screw.	Double rod.
Men's Boot Trees, with 4 feet, - -	$2 25	$3 25	$3 50

	Plain.	Single rod.
Boys' Boot Trees, - - - - - - -	$2 00	$3 00
Gaiter Trees, - - - - - - - -	$3 00 per set.	
Clamps, with Lever, - - - - - - -	65 cts.	
Plain Clamp, - - - - - - - - -	40 "	

CRIMPING BOARDS AND STRETCHERS.

Crimping Boards, plain or for screw, - - - -	Per pair, 40 cts.
Crimp Screws, - - - - - - - -	" 35 cts.
Toe Stretchers, wood - - - - - - - -	$1 00
Instep Stretchers, wood, - - - - - - -	50
Toe Stretchers, iron, - - - - - - - -	1 60
Instep Stretcher, - - - - - - - -	70

SHOE THREADS.

Barbour's Irish Flax Shoe Thread,	No. 10,	- -	Per lb., $	87
Barbour's Irish Flax Shoe Thread,	3,	- - -	"	1 00
Barbour's Irish Flax Shoe Thread,	12,	- -	"	1 30
Barbour's Yellow Stitching Shoe Thread,	12,	- - -	"	1 30
Barbour's 3-Cord Stitching Shoe Thread,	35,	- -	"	1 45
Ullathorne's Yellow Stitching Shoe Thread,	12,	- - -	"	1 30
19 X, Yellow,	- - - -	- -	Per doz.,	60
Smith's Andover Flax Shoe Thread,	No. 10,	- -	- Per lb.,	68
Smith's Andover Flax Shoe Thread,	" 3,	- -	"	84
Smith's Andover Flax Shoe Thread,	" 12,	- - -	"	88

ONEIDA MACHINE SILK.

REDUCED IN PRICE.

We can recommend this silk; we use it ourselves. We don't keep inferior silk.

Black, - - - - -	Per lb., $8 50	Per spool, 80 cents.
White and Orange, full weight, -	" 10 00	" 90 "

Marshall's Imported Machine Thread.

No. 35	40	50	60	70	80	
$1 55	$1 80	$2 20	$2 50	$2 80	$3 00	per lb.
20	25	30	35	40	40	per spool, 2 ounces.

MACHINE NEEDLES.

Howe and Weed Needles, all sizes, - - - 40 cents per dozen.

LASTING OR SERGE.

English Serge, 18 Thread, - - - - - $1 10 per yard.

SHOE BRUSHES.

No. 1,	2	3	4	5	6	8	9
$1 25	$1 75	$2 65	$3 40	$3 50	$3 75	$4 75	$5 50

SHOE LACES.

	$\frac{1}{2}$	$\frac{3}{4}$	$\frac{4}{4}$	$\frac{5}{4}$	$\frac{6}{4}$	$\frac{7}{4}$	$\frac{8}{4}$
Round Cotton, Per gross,	13c.	16c.	20c.	25c.	30c.		
Round Glace Braid, "	15	20	25	30	35	40c.	45c.
Round Ex. qual. Braid, "	25	30	35	40	50	55	65
Flat Glace Braid, "	20	25	32	40	45	50	60
Flat Ex. qual. Braid, "	30	35	45	55	65	75	80
Ex. wide, h'y Flat Braid, "	55	75	90	$1 10	$1 25	$1 40	$1 90

OXFORD TIE LACES.

	$\frac{1}{2}$	$\frac{5}{8}$	$\frac{3}{4}$
Wide, Mohair, – – – – Per gross,	$1 25	$1 50	$1 75
Wide, Alma, – – – – " "	1 75	2 25	2 50

LEATHER LACES.

All lengths, $\frac{4}{4}$ $\frac{5}{4}$ $\frac{6}{4}$ –, – – – – 60 cents per bundle.
100 yards of Lace in each bunch.

BOOT WEBS.

Annaquatuckett, Per Gross, – – – – – – –	$4 00
Anchor, – – – – – – – – – – –	3 90
Extra Colored, – – – – – – – – –	2 90
Round Edge, – – – – – – – – –	1 80
Centennial, – – – – – – – – – –	4 25
Best English, – – – – – – – – –	5 50
Extra Corded, – – – – – – – – –	4 00
¾ Best Gaiter Web, Per Gross, – – – – –	2 00
¾ Gaiter Web, – – – – – – – – –	1 75
¾ Common Gaiter, – – – – – – – –	1 40
¾ French Linen, – – – – – – – – –	1 10

It's the use of tobacco in large quantities that is injurious. Take, for instance, Mr. James Tucker, of Greyson County, Ky., who had a whole hogshead of it fall on him and kill him the other day.

SHOE BLACKING.

	1	2	3	4	5	
French Blacking, Jacquand, Pere & Fils,		30c.	50c.	70c.	85c.	pr dz.
French Blacking, Frank Miller's,	-		65	85		"
Bixby's Best Blacking, - - -	35	40	65	75		"
Mason's Challenge Blacking, -		30	40	60		"
Frank Miller's Water-proof, -●	$1 00	$1 50				

SHOE DRESSINGS.

French Dressing, Brown's, - - - - - -	90c per doz.
Satin Polish, - - - - - - - -	90c. "
Crown Dressing, - - - - - - - -	90 "
Bronze Dressing, Cahills, - - - - - -	$1 90 "
Boston Dressing Blacking, Hauthaway's or Whittemore's,	
quart cans, - - - - - - -	30 cents per can.
Ink, Packard's, - -. - Pints, $1 30	Quarts, $2 00 per doz.
Ink, Packard's, - - - - - -	12 and 20 cents a bottle.
Oven's Varnish, - - - - - - -	25 cents per bottle.
Whittemore's Burnishing Ink, - - Pints, $1 25	Quarts, $2 00

SHOE KNIVES AND RASPS.

Moran & Fulton's No. 0 to 3, - - - - -	Per dozen,	$0 90
Harrington's, - - - - - - - -	"	1 75
Wood's, - - - - - - - -	"	1 40
Wilson's, - - -, - - - - -	"	1 30
Shoe Knives, single, - - - - - -	from 10 to 25 cents.	
French Knives, - - - - - - -	" 40 to	"
Skiving Knives, - - - - - - -	" 20 to 35	"
Patent Extension Knives, - - - - - -	- 75	"

Barnett's or J. Barton Smith's $\frac{1}{4}, \frac{1}{2}, \frac{3}{4}$ File, Rasps, - -

	8in.	9in.	10in.	11in.	12in.
	25c.	30c.	40c.	50c.	60c.
Kit Files, - - - - - - - -	$1 10 per dozen.				

A grocer being solicited to contribute to the building of a church, promptly subscribed his name to the paper in the following eccentric manner : " John Jones (the only place in town where you can get eleven pounds of sugar for a dollar), 25 cents.

SHOE FINDINGS.

NAILS REDUCED.

Fields' Shoe Nails, per ℔., -	6c.	Round Heads,	- -	10c.
Hob Nails, - - -	- 10c.	Zinc Nails,	- -	12c.
Sweedes, - - -	- 11c.	Copper Nails,	- -	42c.

	$\frac{4}{8}$	$\frac{5}{8}$	$\frac{6}{8}$	
Fields' Steel Nails, per M, - -	30c.	35c.	40c.	

	$\frac{5}{8}$	$\frac{4\frac{1}{2}}{8}$	$\frac{4}{8}$	$\frac{3\frac{1}{2}}{8}$	$\frac{3}{8}$
Fields' Channel Nails, per ℔., -	- 25c.	30c.	35c.	40c.	45c.

	3 oz.	2½ oz.	2 oz.	1½ oz.	1 oz.
Fields' Shoe Tacks, - - -	30c.	32c.	37c.	40c.	50c.

Lasting Tacks, per gro., assorted, - 25c. 1 in., 30c. 1¼, 40c. 1½, 50c.
Patent Peg Awls, regular, assorted, - - - - 60 cts. per gro.
Sewing Awls, regular, assorted, - - $1 75 per gro. 18 cts. per doz.
Stitching Awls, regular, assorted, - 1 75 " 18 " "
Russia Bristles, - - - - - - per ℔., $12 00 and $15 00
Russia Bristles, - - - - - per oz., 85 " 1 00
Brass Heel Plates, for ladies, - - - per doz., 65 " 75
Leather Cement, - - - - per bottle, 10 cts. per doz., $1 00
Rubber Cement, - - - - " 15 " " 1 50
Steel Shanks, Gents', R. & L., - - - - - per doz., 25 cts.
Steel Shanks, Gents', straight, - - - - - " 15 "
Steel Shanks, Ladies', straight, - - - - " 10 "
Shoe Lifts, brass, - - - - - Large, $1 25. Small, $1 00
Japan Lifts, - - - - - - - - - 50 cts.
Shoe Lifts, horn, - - - - - - - - - $1 00

Mr. Collins, of Hartford, bought him a dog—a large bloodthirsty bull-dog. He said he wanted a dog that would stand by his wife when he was away. He was away the next night, and came home late, drunk as usual, when his new dog met him at the door. The dog looked at him as much as to say, "Go back where you got your whisky." Mr. Collins argued with the dog, asking him who was bossing the house. The dog took a mouthful out of Mr. Collins, right where he didn't want it taken out, and the owner of the dog is now troubled about sitting down. He wants to sell a good dog.

SHOE FINDINGS.

Cork Soles, Gents', - - - - - - -	80 cents per doz.
Cork Soles, Ladies', - - - - - -	70 " "
Cork Soles, 3-16th for cork sole boots, - -	$1 50 " "
Corks for Lifts, according to thickness, - -	25 cents to $1 00 each.
Gum Tragacanth, - - - - - - - -	" $1 00
Gum Arabic, - - - - - - - -	" 35 cts.
French Chalk, in paper, - - - - - - -	" 10 "
French Chalk, in tin cans, - -	1 lb., 15c. 2 lb., 25c. 5 lb., 50c.
Best Clarified Wax, - - - - - - - -	50c. per 100.
Shoe Makers' Hanging Signs, - - - - - -	$1 25 each
Shoe Makers' Benches, - - - - - - -	90 "
Soap Stones, - - - - - - - - -	12 "
Sand Stones, - -	SMALL, 8 cts.; MEDIUM, 12 cts.; LARGE, 15 cts.
Sand Paper, - - - - - - - - -	per doz., 12 "
Emery Paper, - - - - - - - -	" 25 "
Size Sticks, - PLAIN, 18 cts.; SINGLE FOLD'G, 35 cts; ENTIRE FOLD'G, 45 "	
Heel Ball, - - - - - -	No. 1, 10 cts.; No. 3, 30 "
Heel Ball, in long sticks, - - - - -	40 cts. per doz.
Buttons for Men's Gaiters, - - - - -	per gross, 15c.
Buttons for Men's Gaiters, best, - - - - - -	" 30c.
Patent Buttons, No. 3, men's, - - - - -	" $3 80
Patent Buttons, No. 2, ladies', - - - - -	" 3 70
Patent Rivets, - - - - - - -	per lb., 90
Machine for Buttons, - - - - - - -	each $1 75
Button Hooks, (all qualities,) - - -	from $1 00 to $4 00 per gross
Patent Pegging Hafts, - - - -	10 cts. each. $1 00 per doz.
Sewing and Pegging Hafts, - - - - -	18 cts. "
Patent Sewing Hafts, - - - -	15 cts. each. $1 50 "
Eyelets, B, Long, Black or White, - - -	1,000 Box, 13 cts.
" " " " " " - - -	10,000 Box, $1 10
Shoe Pegs, in half Bushel Bags, 60c. each, - -	Per Bushel, $1 15

An Irish paper published the following item: "A deaf man named Taff was run down by a passenger train, and killed, on Wednesday morning. He was injured in a similar way about a year ago."

Turner, the painter, was at a dinner where several artists, amateurs, and literary men were convened. A poet, by way of being facetious, proposed as a toast, "The painters and glaziers of England." The toast was drunk; and Turner, after returning thanks for it, proposed, "Success to the paper-stainers," and called on the poet to respond.

CRIMPING MACHINES.

Brass Jaw,	Each	$18 00
Buffalo Crimper,	Each	9 00

IRON STANDS.

Counter Stand, 3 feet,	$ 85
Counter Stand, extra high, 3 feet,	1 25
Improved Iron Stand, latest style,	1 25
Splitting Machine,	8 in., $2 50 ; 10 in., $3 00 ; 12 in., $3 50

BOOT SIGNS.

Hollow Zinc Boot Leg Signs, 2 feet high, Painted,	$6 00
" " " " " " " Gold Gilt,	8 00

Packed in Box for shipping, 50 cents extra.

☞ This is the handsomest sign ever made for the shoe business.

HOOKS.

$1.90

per Thousand.

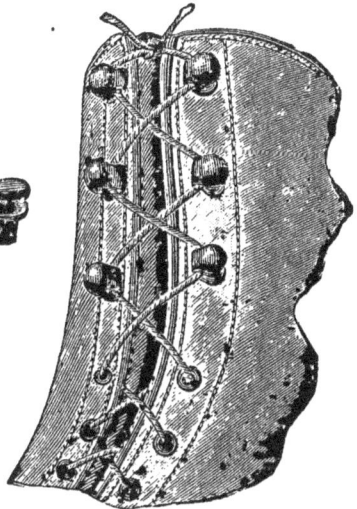

HOOK MACHINES.

Small Size,	$2.50
Large "	5.00

500 HOOKS IN A BOX.

SHOE TOOLS.

Double Colice,	-	-	30 cts.	Heel Shaves,	-	-	65 cts.
Single Colice,	-	-	25	Edge Shaves,	-	-	-	75
Double Shank Iron,	-	25	Edge Plane,	-	-	-	50
Single Shank Iron,	-	-	20	Long Sticks,	-	-	-	30
Top Channel Set,	-	-	30	Colted Stick,	-	-	45
Heel Burnisher,	-	-	-	30 cts.	Shoulder Stick,	-	-	25
Corrugated Burnisher,	-	50	Emery Knife Strap,	-	20 cts.
Shank Burnisher,	-	-	40	Scratch and Slick Bones,	-	15
Rhan Break and Key,	-	45	Buffers,	-	-	-	10
Rhan File,	-	-	-	15	Compasses,	-	-	-	15
Jigger, long handle,	-	-	30 cts.	Clamming Markers,	-	25
Jigger, short handle,	-	-	25	Yankee Cutter, long handle, 35
Welt Knife,	-	-	-	15	Yankee Cutter, short handle, 25 cts.
Welt Mill,	-	-	-	-	10	Plain Peg Break,	-	-	40
Welt Trimmer,	-	-	75	Swivel Peg Break,	-	50
Channel Gouge,	-	-	-	25	Round Heel Cutter,	-	30
Seam Set,	-	-	-	15 cts.	Counter Stand, with 2
Strip Awls,	-	-	-	15	Heel Irons,	-	$1 50
Channel Opener,	-	-	15	Shoe Hammers,	-	25 to 40 cts.
Stitch Divider,	-	-	-	15	Shank Lasters (Crab),	-	35
Last Hook,	-	-	-	15	French Lasters,	-	-	$1 00
Boot Hook,	-	-	-	35	Alligator Boot Jack,	-	20
Boot Hook, bone handle,	-	60 cts.	Eyelet Punch,	-	-	-	50
Peg Wheels and Handle,	-	25	Revolving Punch, 4 sizes, $1 50
Shank Wheel,	-	-	-	25	Eyelet Set,	-	-	-	50 cts.
Fudge Wheel,	-	-	-	30	Box Wheel, with Slide,	-	50
Bottom Wheel,	-	-	25	Box Wheel, Extra,	-	80
Timmon's Pincers	-	40 to 65	French Wheel and Key,	-	40
Laufenberger's "	$1 25 to $1 75	Timmon's Nippers,	-	25
Heel Knives,	-	-	-	25	Steel Nippers,	-	-	-	35

HE WAS MISTAKEN.

This morning a gentleman entered a shop on Fifth street, and asked the clerk, "What is the price of knit undershirts with breast pockets?" He added, "I travel a great deal and carry large amounts of money, and I think that idea of pockets an excellent one, and I am surprised that some one has not thought of it before." "Really sir," replied the clerk, "I think myself it would be a good plan, but I am sorry to say we have none, and I did not know there were any made."

"You did not?" said the customer. "Well that's singular. They are exhibited in your window, and caught my eye as I was passing." "You must be mistaken," said the clerk. "I know every article in the store."

"But I am not," persisted the gentleman. "Step around and see for yourself." The wondering shopkeeper did as requested. He stepped briskly to the front of the window, looked in, then looked at the gentleman, then coughed, and acted as though he had just felt a sudden pain in the stomach, and then rammed a handkerchief into his mouth, and stepped back behind the counter.

"Well," exclaimed the customer triumphantly, "ain't they there?" "Ye—yes," said the clerk, appearing as though he had a fish bone in his throat. "They are there sure enough. But, sir, those undershirts are not for men, and those pockets are —— and at this point he dived under the counter and disappeared, while a young lady clerk, standing near, smothered a convulsive giggle in a cambric handkerchief and started off with a very red face on important business to the rear part of the store. A sudden light seemed to break in upon the stranger, and he departed hurriedly, muttering, "How in thunder could I tell? I ain't a married man, and can't be expected to know everything."

———

Nobody but a fool will spend his time trying to convince a fool.

Courage without discretion is a ram with horns on both ends, he will have more fights than he can well attend to.

Thoughtfulness for others, generosity, modesty and self-respect are the falities which make a real gentleman or lady.

HE WAS A WAYFARER.

He was one of the most hard-up and broken-down individuals you ever saw.

An old bent-up chap, who looked as if he had been struck continually in the back and doubled up by a trip-hammer, and then instantly struck in the stomach and knocked back again by another. And the expression of misery which rested upon his venerable and dusty face would have melted the heart of a statue.

Therefore, when he dropped into a down-town restaurant, and leaned feebly against the cashier's desk, the proprietor beamed on him benignly.

" This is an eatin'-place ?" said he, inquiringly.

" Yes," answered the proprietor, encouragingly.

" A place where you eat ?" continued the old man, pathetically blowing his nose on a rusty coat sleeve.

" Yes."

" And I s'pose you keep everything ?"

" Of course."

" Have waiters ?"

" Certainly'"

" An' put clean white aprons on them every day ?"

" Well," answered the proprietor, thoughtfully, " I can't say as I do change their aprons daily."

The venerable stranger sighed.

" I'm a pilgrim and a wayfarer," groaned he ; " an' it won't be long afore I'll be ridin' in a hearse ; but, young man, if you venerate gray hairs, jest listen to my advice. Allus put clean white aprons on your waiters daily."

The restaurant keeper appeared suitably impressed, and said he should think about it.

" That's right," said the wayfarer ; " 'heed to the words of your forefathers,' the Scripter says. Now, young man, to look at me, you wouldn't think that I was worth a million."

" No."

" Or that I owned the Erie Railroad and the Cunard line ?"

" No."

" Or that on Texas prairies I hev ten thousand cattle a-tearin' an' a-gambolin' around in their innocent freedom, with my monogram stamped onto their hind quarters ?"

The astonished proprietor said that he would never for a moment have imagined such a thing.

" I see," sorrowfully muttered the aged millionaire ; " you took me for a poet or a free-lunch fiend, didn't you ?"

The man of meals at any hour said he suspected something of the sort.

" I knew so," said his visitor ; " thus the world misjudges me. But let me whisper in your ear. I haven't the faintest idea what to do with my immense wealth, for I am alone in this world. So I've decided on a plan. I go into a restaurant, a nice place like this, an' I order a sumptuous meal—boned turkey, quail on toast, lobster salad, an' etcetra. When I finish I walk out. Now, if the young man what keeps it don't run after me, take me by the collar and demand my bill, but lets me go in peace, I come back in a few days in a gorjuss ekipage present that young man with a gold watch an' a set of diamond studs an' a silver-plated revolver. I take that young man, figgeratively speakin', to my buzum, an' treat him like an only son. I surround that young man with luxury, lavish thousands on him, let him go to the theater, or concert, or prayer meetin', a funeral or any other place where he can enjoy himself, every night if he wishes to. I allow him two servants, a bath room with hot and cold water, and give him a new pair of suspenders every Christmas. An' finally, when I go to join the heavenly band, I leave him all my money. Do you see ?"

" Do I see ?" echoed the young man ; " yes, I see too muchly. Just light out, old sport, that dodge won't work here."

" Sir," haughtily uttered the aged philanthropist, " what do you mean ?"

" I mean," answered the other, " that you don't get no free meals here. Now climb out, old Peter Cooper, before I bounce you."

" How I was mistaken," groaned the owner of the Cunard line, moving away. " Young man, I thought I saw in you a righteous youth ; but 'tis not so. Evil is marked in indelible ink upon your brow, and when you stand upon the gallows I won't move a finger to help you."

" Get out, will you ?"

The philanthropist paused a moment, waved his umbrella impressively in the air, and then said, as he resumed his journey :

" Look in the morning papers, to-morrow. I owe a duty to society, and I shall perform it. The public, sir, shall know that you put dog's meat into your sausage, and cat's flesh into your hash. Beware, sir, beware !"

And before the proprietor of the place could reach a club and climb over the counter, he was gone.

THE GAME OF LIFE.

Man's life is a game of cards. First, it is "crib-age," next he tries to "go it alone," at a sort of "cut, shuffle and deal" pace. Then he "gambols on the green." Then he "raises" the "deuce" when his mother "takes a hand in," and, contrary to Hoyle, "beats the little joker" wih her "five." Then, with his "diamonds" he "wins" the "queen of hearts." Tired of "playing a lone hand," he expresses a desire to "assist" his fair "partner," "throws out his cards," and the clergyman takes a ten-dollar bill out of him "on a pair." She "orders him up" to build fires. Like a "knave" he joins the "clubs," where he often gets "high," which is "low," too. If he keeps "straight" he is oftentimes "flush." He grows old and "bluff," sees a "deal" of trouble, when at last he "shuffles" off his mortal coil, and "passes in his checks," as he is "raked in" by a "spade," life's fitful "game" is ended, and he waits the summons of Gabriel's "trump," which shall "order him up."

GOOD TASTE.

"I can't bear children," said Miss Prim, disdainfully.

Mrs. Partington looked at her over her spectacles, mildly, before she replied:

"Perhaps if you could you would like them better."